I0482572

Time

Michael Yeow

About the author

I am not an expert in the field, but have acquired knowledge in Quantum Mechanics (2005) and Cosmology (2009) through studying with the Open University. I truly believe that all disciplines of knowledge in the modern world came from the books salvaged from the Fire on the Christmas' Eve in 1824. I could only conjecture about this but only the true ancients can verify this fact.

Recently I had been detained for seven months under the British Mental Health Act Section Three for some unclear reasons (or that was how I perceived the whole thing).

In this little book, I want to share with you how I see the puzzling questions that have bamboozled even the best minds in physics.

Preface

Quantum physicists did many experiments verifying the puzzling features of the Quantum World. Those salvaged readily perusable physics books all told of these strange incredible aspects of quantum physics. As a consequence of the repeated experiments, all quantum physicists had a consensual view on the subject even though they don't totally believe in it – just like when Albert Einstein uttered, "God does not play dice with the Universe."

What is time? Scientists define time as the fourth dimension but, strictly speaking, time is just a tag in this 3-D space, not unlike the clock hanging on the wall telling you the time in this 3-D room that it is twelve o'clock midday! Time has no unit of length and so whether it is 4-D spacetime or 3-D space, the boundary of the *space* concerned is still the same but in the 4-D world we have time tagging along!

Our physical reality is this 4-D Universe, with space and time thrown-in. In which, we have the past, the present and the future stories. This intrinsically implies that the arrow of time is unidirectional or one-way forward. Why? The constraining factor that causes this is the speed of light, i.e. 3×10^8 ms^{-1}. Could this speed limit be broken? Could one really go back in time to kill the grandfather? Could we teleport?

Albert Einstein was acknowledged as the greatest genius in Physics, on par with Stephen Hawking. His most important three credits were:

$$E = m\,c^2$$

Special Relativity

General Relativity

I had mentioned about the household equation in my previous books. It tells of a creation story and not what Einstein had misconstrued.

Special Relativity states that the laws of physics are invariant (i.e. identical) in all inertial systems (non-accelerating frames of reference), and that the speed of light in a vacuum is the same for all observers, regardless of the motion of the light source.

In other words, all inertial observers basically experience the same laws of physics. Dear friends, this means we, all fellow earthlings, face and experience the same physical laws of what life throws at us.

The speed of light in a vacuum is a constant for all observers, regardless of the motion of the light source. At first glance, this seems difficult to understand because it is regardless of relative motion between the light source and the observer. However, the mysterious crunch point is that light is a transverse wave. The buoy out at the sea is bobbing up and down, but the sea waves (transverse waves) keep rushing in! So when you ride on a beam of light, the beam won't take you anywhere (but you would be bobbing up and down only). Every point on the light wave is bobbing up and down at right angles to the direction of the travelling light wave, but every point remains at its spot. There is no parallel relative motion at all! This makes the speed of light independent of the relative motion between the light source and the observer.

We all heard of the duality of light, that is, light is both a wave and a particle. This is simple too. Yes, the light is a transverse wave. Many transverse waves add up to give wave packets, which are particles.

c is the speed of light, f is the frequency of light and λ is the wavelength of light,

$$c = f\lambda$$

this means c is the product of f and λ and, because c is a constant this means that when f increases λ will decrease and vice versa.

Extremely low frequency (ELF) is 3 Hz in frequency, 100Mm in wavelength and 12.4feV in energy.

Gamma rays is 300 EHz in frequency, 1pm in wavelength and 1.24MeV in energy.

The energy is dependent on the frequency: higher frequency gives higher energy. We have higher energy at the higher spectrum of light. As mass could interchange with energy, higher energy means higher mass. That is the reason why when we approach the speed of light, length (wavelength) shortens and mass increases.

General Relativity is essentially the Equivalence Principle in action. Basically, in the presence of great mass, there is no difference between acceleration and gravity. In addition, gravity can deflect light and generate e.g. 'gravitational lenses' etc.

My reasoning is that if gravity could have an effect on light and this must therefore indicate light has mass. Gravity is due to the (love) nature of mass and, hence two masses will attract. I believe when Eddington Solar Eclipse Experiment confirmed with Einstein's prediction, the light beam was deflected towards the Sun rather than away from it.

The arrow of time is one-way and it is in forward direction. How do we travel back in time? Easy! Just take a walk down the memory lane and you could achieve that easily if you don't have a memory problem.

The temporal part of our brains let us have the framework and perspectives of time and its passage. We have media like photos, audio/video recordings to help us store our memories.

"We can't change our past, but we could easily change our future."

No. The past has passed, and we can't change our past. We could not go back in time to kill the grandfather because we weren't born yet. However, we could change our future. If you don't like the sound of your fortuneteller, then take action and alter the course of your life journey to rewrite yourself a beautiful future the way you

have wanted it.

Quantum physicists were fascinated by the no cloning phenomenon and complete randomness of pre-measured photons in certain polarization/ coding experiments. Yes, even a speck has a choice of freewill!

Some ancients may have experienced time travel and teleportation. These two novelties had been the foretaste from the Creator.

The program codes intended for time travel, teleportation and superluminal speed are not fully completed by the Creator yet...

When He returns, we'll be thrilled!